Beautiful Birds

I0799599

OSTRICHES

By Bray Jacobson

Gareth Stevens
PUBLISHING

Please visit our website, www.garethstevens.com. For a free color catalog of all our high-quality books, call toll free 1-800-542-2595 or fax 1-877-542-2596.

Library of Congress Cataloging-in-Publication Data

Names: Jacobson, Bray, author.
Title: Ostriches / Bray Jacobson.
Description: Buffalo, New York : Gareth Stevens Publishing, 2024. | Series: Beautiful birds | Includes index.
Identifiers: LCCN 2023023714 | ISBN 9781538292532 (library binding) | ISBN 9781538292525 (paperback) | ISBN 9781538292549 (ebook)
Subjects: LCSH: Ostriches–Juvenile literature.
Classification: LCC QL696.S9 J26 2024 | DDC 598.5/24–dc23/eng/20230526
LC record available at https://lccn.loc.gov/2023023714

First Edition

Published in 2024 by
Gareth Stevens Publishing
2544 Clinton St.
Buffalo, NY 14224

Editor: Kristen Nelson
Designer: Andrea Davison-Bartolotta

Photo credits: Cover, p. 21 Roger de la Harpe/Shutterstock.com; p. 1 SABPICS/Shutterstock.com; p. 5 Sergei25/Shutterstock.com; p. 7 Lars Fortuin/Shutterstock.com; p. 9 Art Konovalov/Shutterstock.com; pp. 11, 24 (wings) danielecolombo/Shutterstock.com; pp. 13, 24 (claw) Nikolai Link/Shutterstock.com; p. 15 wayak/Shutterstock.com; p. 17 Photos314159/Shutterstock.com; p. 19 Goldilock Project/Shutterstock.com; pp. 23, 24 (chick) Dmussman/Shutterstock.com.

Printed in the United States of America

CPSIA compliance information: Batch #CW24GS: For further information contact Gareth Stevens, New York, New York at 1-800-542-2595.

Contents

Where on Earth? 4

Ostrich Bodies 8

Ostrich Life 14

Nesting 20

Words to Know 24

Index. 24

An ostrich is a bird.
It is the biggest bird
on Earth!

It only lives in Africa.

It can be black
and white.
It can be gray
and brown.

It has wings.
It has feathers.
But it does not fly!

It has two feet.
Feet have two toes.
One toe has a claw.

It runs fast.
It kicks hard!

It mostly eats plants.
It may eat bugs and
snakes too.

It may live in a group.
Mothers in a group
share a nest.

Mothers lay large eggs.
No other bird's
are bigger!

Babies are chicks.
They grow fast.

Words to Know

chick

claw

wings

Index

Africa, 6
body, 10, 12
colors, 8
food, 16